GW01142533

Dutch ambulances (1945-1975)

by K.J.J. Waldeck, M.D., Ph.D.

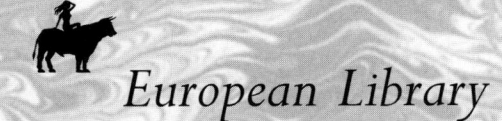

European Library   ZALTBOMMEL/THE NETHERLANDS

Cover:
Packard (1949) of the Municipal Health Service (GGD) of Amsterdam, built in larger numbers by Akkermans from Oud Gastel. It was a real miracle these vehicles, decorated with a lot of chromium, could be used as an ambulance only a few year after World War II.

Original title: Ambulances in beeld (1945-1975). Van ziekenwagen tot ambulance.
Translation: Anton L. Hermans.

BACK IN TIME

GB ISBN 90 288 2043 4
© 2000 European Library – Zaltbommel/The Netherlands

No part of this book may be reproduced in any form, by print, photoprint, microfilm or any other means, without written permission from the publisher.

European Library
post office box 49
NL – 5300 AA Zaltbommel/The Netherlands
telephone: +31 418 513144
fax: +31 418 515515
e-mail:publisher@eurobib.nl

Introduction

Ambulances survived the war but they were hardly usable anymore to fulfil their purpose. In the reconstruction period after World War II the transportation of the seriously ill and injured had to be restored therefore as well.

The patients

As a result of the war, complicated and radical medicamental as well as surgical treatments became available, namely antibiotics for severe infections and surgery of the lung for tuberculosis. These conditions could only be treated in hospitals. The ill and injured usually were not at the top of their form when they went to the hospital and therefore they had to be transported recumbently. The most likely vehicle for that purpose after the war was to be the ambulance, despite the shortage of chassis to built these vehicles on.

Care and care providers

Up to twenty-five years after the war the aid to patients was still confined to bandaging, putting splints on, and the transportation of seriously ill and injured people. The successfully life saving methods, like mouth-to-mouth resuscitation and external cardiac massage, were not to be rediscovered until the late fifties. At the well-known services the ambulance crew consisted of a driver, who was also the stretcher-bearer, and an attendant. These professionals did not have a specific training for their job, but much was learnt by experience. Outside the large cities it was not an uncommon view that the crew consisted of only the driver/ stretcher-bearer who was merely a first-aider. The daily job of the driver could for instance be a motor mechanic or even a pall-bearer for an undertaker. It also occurred that local cab drivers, bakers or butchers took place behind the wheel if that was necessary. Suchlike drivers could not do more than open the oxygen device, which was fitted in the dashboard of the vehicle, a little bit more if they heard the patient moan.

Equipment and vehicles

The equipment was very sober until the end of the sixties. On board there were enough cloth and blankets to wrap the patients up sufficiently. The aids were restricted to a urinal, a chamber pot, and some small sandbags, which were used to support broken limbs. Beside these attributes there would normally be a first-aid kit so that a bandage could be applied if injuries were sustained and a few wire splints. At some larger, local services the so-called Pulmomat® was placed in the vehicle during the poliomyelitis epidemic in the fifties. Since then the crew was able to perform artificial respiration at people who were suffering from breath paralysis. The equipment was not suitable on the other hand to put patients on life support who suffered from an acute illness or an injury of the breast after an accident. The rediscovery of the mouth-to-mouth resuscitation and external cardiac massage opened the possibility of direct medical attendance outside the hospital. It was only after this rediscovery that suction and oxygen apparatus were fitted in the ambulances. In later years more adequate respirators made their appearance.

After the war the ambulance services had only a few vehicles at their disposal; namely the ambulances from before the war and the few military ambulances of the Allies which survived, like Austins, Fords, and Chevrolets. In order to transport civilians the chassis of these old ambulances were used to build new ambulances on. When the Marshall help of 1947 had been well under way it gradually became possible to import chassis of American passenger cars. Dutch coachbuilders like Visser Bros. from Leeuwarden, and Akkermans from Oud Gastel, built unique ambulance coaches on those American chassis. In the sixties these vehicles were replaced by German ambulances. Binz from Lorch and Miesen situated in Bonn (former West-Germany) made these more or less standardised ambulances. The American ambulances were hard to forget and that is why especially Chevrolets with a Dutch body remained very popular.

Organisation

Until the 1970s the ambulance service was controlled by local health services, privately owned businesses, hospitals, and to a small extent by the Red Cross and some first-aid associations. There was hardly any organisation. Anyone who wanted to establish an ambulance service could do so as there were no real demands that had to be met. Even though there was no national emergency phone number, each local council did have its own emergency number. The station on the other hand was not permanently staffed which resulted in enough ambulances from all four quarters of the compass at the place of the accident, or not even one from the local region. Much discussion took place concerning the transportation of the patient to a hospital, even at the scene of the accident in front of the patient. The national government did not stay passive concerning the developments in the area of urgent medical attention. The government took the initiative in a bill, called the Ambulance Services Act (Wet Ambulancevervoer), which became law in 1971. The law was revised in 1976. It still was a general Act that was to be followed by further decrees in the years to follow like the decree of Requirements of Ambulance services (Eisenbesluit Ambulancevervoer) from 1976 and the 1978 decree on the Inventory of Ambulance cars (Inventarisbesluit Ambulancevervoer). The ambulance had been transformed into a Mobile Intensive Care Unit. In this unit all the absolutely necessary medical attention could be given to the patient without having to wait until the patient was actually in the hospital. Finally justice had been done to the word ambulance. It regained its former glory of 'hôpital ambulante' or in other words: mobile hospital. The decree of Requirements of Ambulance services (Eisenbesluit Ambulancevervoer) set the minimal requirements an ambulance had to meet. Especially the equipment of the ambulance, and full knowledge and skill of the ambulance crew were the main aspects that would change after 1975.

1   Austin K2 Y, 2 tons, 4 x 2 (approx. 1942) of the Central Board of the Dutch Red Cross. A large number of these former British army ambulances were donated to the Dutch after World War II. These Austins are lined up in the compound of the Red Cross on Fahrenheitstraat, at the corner of Laan van Meerdervoort in The Hague. The original colour of these vehicles had been army green. In a later stadium many were sprayed white.

2   Austin K2Y, 2 tons, 4 x 2 (about 1942) of the Central Board of the Dutch Red Cross. A permanent door has replaced the original canvas door. The registration number is currently a civilian number in South-Holland. The sign of the Red Cross is still a small one as in wartime the cross was an easily target to aim at.

3 Austin K2Y, 2 tons, 4 x 2 (circa 1942) of the Central Board of the Dutch Red Cross. In the garage of the Dutch Red Cross on Fahrenheitstraat, at the corner of Laan van Meerdervoort in The Hague all the Austins were sprayed white. The sign of the Red Cross had been increased compared to the ones used during war. The chance that another shot would be fired at the vehicle because of the cross was highly unlikely. After the war these Austins were brought into action in case of a disaster.

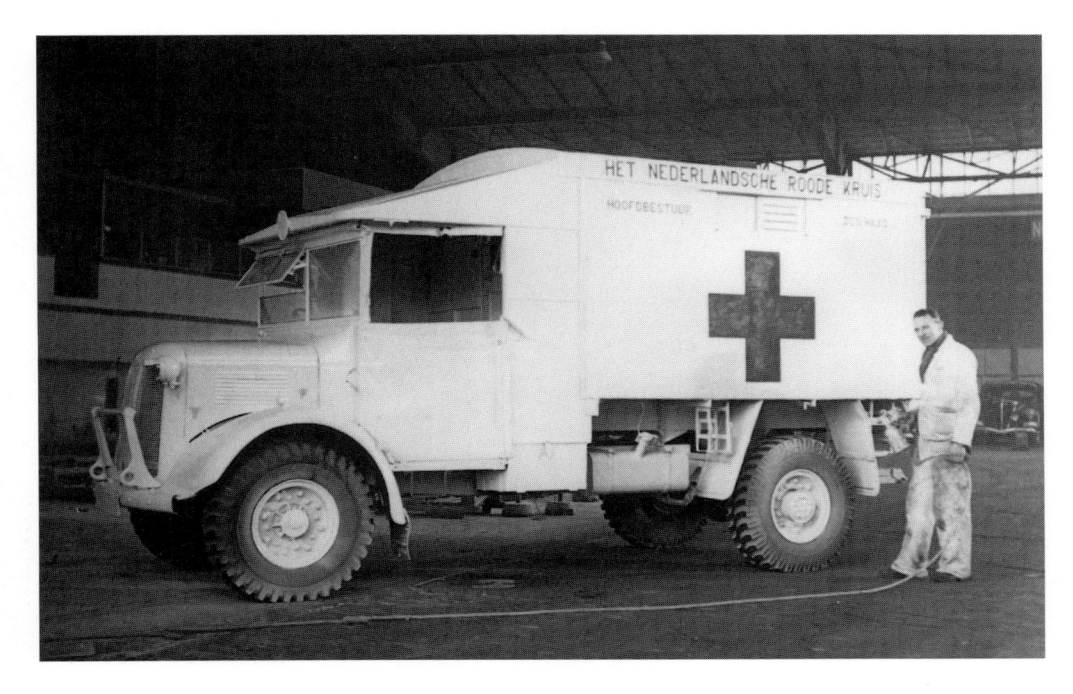

4   Austin K2Y, 2 tons, 4 x 2 (approx. 1942) of the Rotterdam delegation of the Dutch Red Cross with its original canvas door. In this model a maximum of four patients could be transported at the same time. The Royal Dutch Airlines (KLM) also used this type at Amsterdam Airport.

5   Left: Austin K2 Y, 2 tons, 4 x 2 (approx. 1942) built by Visser Bros., Leeuwarden on the frame of an army ambulance. Probably of the Dutch Red Cross because of the white trim and the registration number. The Municipal Health Service (GGD) in Arnhem used a similar kind of Austin.

The right picture shows the same Austin K2 Y, 2 tons, 4 x 2 (approx. 1948). The interior is Spartan-like and is compensated by the leaded windows with the Red Cross sign.

6 Chrysler (1948) of one of the private ambulance services (Haagsche Ziekenvervoer Onderneming) of The Hague. The company was owned by the brother of the professor in surgery at Groningen, but born in The Hague, Michaël. The stretcher had to be pushed in through the rear of the vehicle. It would not be possible for a pregnant woman with a bulged abdomen to enter the ambulance lying on a stretcher. Similar cars (such as De Sotos) were used by the first private ambulance service (Eerste Particuliere Model Ambulance), Amsterdam.

7   Ford Fordor (1949) of the private firm Vreeling & Troost, built by Visser Bros. from Leeuwarden. The registration number dates from 1951. This Ford model had been imported in 1949 and could be imported as a result of the Marshall help of 1948. The vehicle has a firm line and had no rotating lights, mirrors or any other attributes on the exterior.

8 Left: Packard (1950) of the Dutch Red Cross. This picture has probably been taken on the promenade of Scheveningen. The body-work came from Visser Bros., Leeuwarden. The huge front of the car has been decorated with a swan. In after years these ornaments were banned because of the potential danger they could have for pedestrians and cyclists in case of a collision.

Right: Packard (1950). The model of the stretcher is one of the Riemvis-type with big, rounded suspension-feathers. Next to the stretcher a so-called sedan chair.

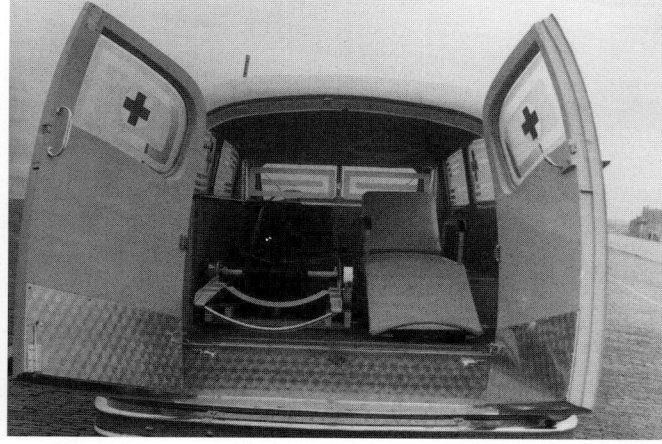

9   Commer (about.1950). For the Netherlands an extraordinary ambulance built by Visser Bros., Leeuwarden. It is not known who ordered this vehicle. It had been built on a frame of a delivery van. It is also unknown what the purpose of this ambulance on a van-chassis was.

10 Packard (1952) of the Municipal Health Service (G.G. & G.D.), Amsterdam. Built by Akkermans, Oud Gastel. The frame of the Packard was much sought in the first five or ten years after the war. On these chassis ambulances were built for the municipal public health departments. These impressive ambulances could easily transport two patients simultaneously. The front of the ambulance is not as impressing as its predecessor (see cover).

11   Studebaker (circa 1951) of the First Aid Service (Eerste Hulpdienst, EHD) of the Municipal Health Service (GGD) of Leiden, delivered by the firm of Boon, Leiden. This photo had been taken on the grounds of the fire department at Langebrug when the ambulance entered into service. Next to the Studebaker stands an A-Ford, which survived the Second World War. This picture is the first of the author's collection. He started his collection when he was still a medical student and worked for the EHD on a regular basis as a male nurse.

12 Austin A70 (1953) of the private ambulance service Veenstra Ziekenvervoer from Amsterdam, built by Visser Bros., Leeuwarden. At least two other examples had been built. That model was intended for Indonesia (Rumah Sakit [hospital] Brajat Minulja in Surakarta (Java).

13 Ford Customline (1953) of the Municipal Health Service of Haarlem, built by Visser Bros., Leeuwarden. Haarlem had the disposal of an 'accident service' (ongevallendienst) or in brief O.D. since 1927. Qualified members of various first aid associations took part in this O.D.. The council paid the telephone connections of a dozen members, who lived across the city. These people could be called upon in case of an emergency. Also 15 oxygen apparatus had been placed in the city. Only in case of a serious accident the ambulance would be called in, next to the nearest member of the O.D. and if necessary a doctor.

14 Ford F250 (1953) of the Municipal Health Service (GGD) of Amersfoort, built by Visser Bros., Leeuwarden. At the beginning of the 1950s this robust ambulances had been built on as well small lorry chassis as on chassis of the Studebaker, Chevrolet, and Ford. This picture had been taken in front of Gosse Visser's at the P.C.Hooftstraat in Leeuwarden. Gosse is the eldest of the four Visser brothers and still lives there.

15   Studebaker (1953) of the Municipal Health Service (Geneeskundige Dienst), Nijmegen. Several coachbuilders built ambulances for municipal services on the Studebaker chassis. This bodywork is from Vermeulen, Haarlem. Later, Vermeulen gained more reputation as manufacturer of sunshine roofs for passenger cars.

16 Chevrolet 2100 Two-Ten (1954) of the St. Liduina Hospital, Apeldoorn, built by Visser Bros., Leeuwarden. Many hospitals set themselves as task the transportation of the seriously ill and injured in the first ten years after World War II. The house of Gosse Visser is in the background. In front of the house was a playing yard. The Chevrolet stands almost in front of the domestic photographer of the Visser Bros., Frans Popken, on De Arumerstraat.

17 Chevrolet 3100 (1954) built by the coachbuilder BERWI. BERWI stands for BERgen from WInschoten. A similar kind of car without the Red Cross above the windshield had been found in Leiden in 1971. The original registration number of that ambulance was RG-19-54. Currently it is a student's car. Berwi is up to now well known as manufacturer of fire-engines.

18   Left: Ford Customline (1954) of the Municipal Health Service (GGD) from Tilburg, built by Akkermans, Oud Gastel. This Brabantine coachbuilder had several municipal services as his regular clients, like those of Rotterdam, Breda, and Tilburg.

Right: Ford Customline (1954). This ambulance had a lot of space compared to the models out of the 1960s like the Commers, Ford Transits, and Volkswagen mini vans. The interior was simple and only fitted for the transportation of the seriously ill. Additional springs for the stretcher were not necessary. These American cars had already enough suspension action, so people said.

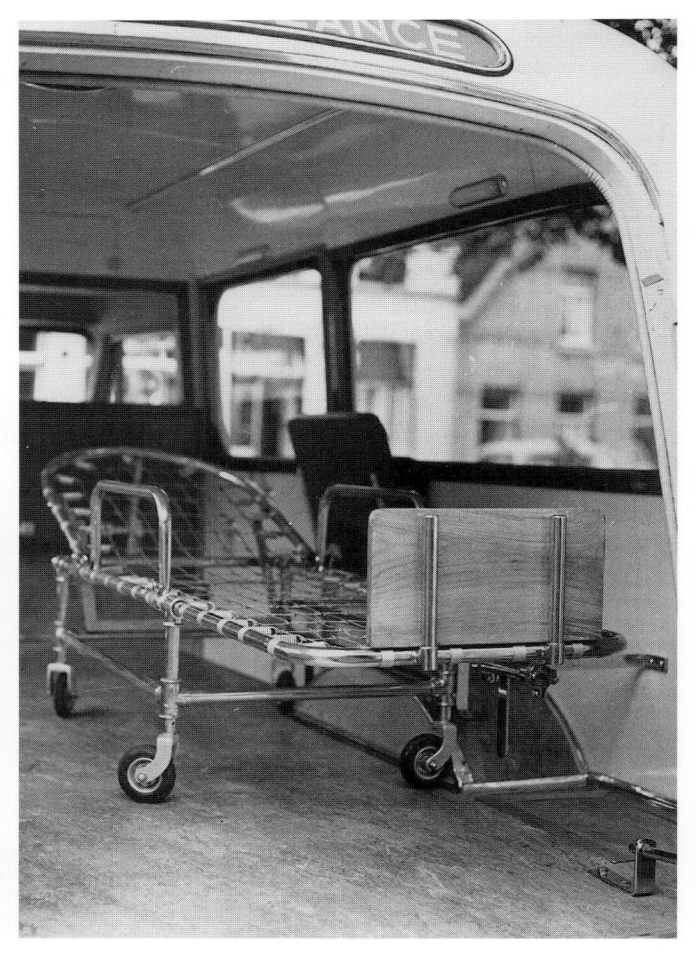

19   Opel Blitz (1955) of the Dutch Red Cross, department Oude IJssel (Doetinchem), built by Miesen from Bonn (then: West-Germany) and delivered by NEDAM, Roermond (The Netherlands). Ambulance services imported vehicles built on light truck Ford and Chevrolet chassis from America. In this period on the other hand also these Opels were ordered for the transportation of the seriously ill and injured.

20 Cadillac Sedan de Ville (1956), built by Visser Bros., Leeuwarden. Suchlike ambulance was the Marilyn Monroe of the ambulances. On the front seat four people could be seated. One of these people sat on the left of the driver. Only the larger communities like Utrecht, The Hague and Amsterdam were able to purchase these lovely vehicles.

21   Chevrolet (1956) of the St. Jansgasthuis from Weert, built by Miesen, Bonn and delivered by NEDAM, Roermond. Ten years after World War II ambulances were built on American chassis in West-Germany meant for the Dutch market. Behind the split side-door on the left the spare wheel and rescue equipment could be found. On Mercedes Benz these doors had been used for many years as well.

22 Ford Taunus Transit (approx. 1956) fitted up as an ambulance by BERWI Coachbuilders, Winschoten. Two patients could be put in on a stretcher in this ambulance. These 'cans' could not weight up to the solid, American ambulances. The high rate of the Dollar and the start of the EEG in 1957 were to be the reasons that the purchase of non-EEG vehicles would become more difficult and more expensive.

23   Mercury Monterey (1956), presumably built by Visser Bros., Leeuwarden. The registration number dates from 1958. It was not uncommon for coachbuilders to use damaged cars, which had been declared as 'total loss', to built ambulances on. In many cases the plating and the chassis had to be restored. Thereby the chassis needed to be extended. That is why the type of a car is, in some cases, dated a few years earlier than the registration makes us believe. During the oil crisis a LPG-tank had been installed under this ambulance.

24 Ford Taunus Transit (1957) of the medical service (medische dienst, GED) of the Amsterdam Droogdok Maatschappij (shipbuilders), fitted up as ambulance by Visser Bros., Leeuwarden. Several coach builders in and outside the Netherlands made dozens of these little Fords suitable for the transportation of the seriously ill and injured. These ambulances were relatively cheap as the outside body already existed. Only the interior required some attention.

25 Left: Ford Fairlane 500 (1957) of the Municipal Health Service (GGD) of Tilburg, built by Akkermans, Oud Gastel. Of the original chassis and bodywork was not much left. Just the front of the car was a recognisable feature. Chassis were often extended. The rear lights originate to a Mercedes-Benz from those years. The most striking feature is that the backdoor opened from the left instead of the other way round. The ambulances drove round with the coat of arms of the city on the doors. There was not yet a sign of the 'Star of Life' and other special striping.

Right: Ford Fairlane 500 (1957). The impressive the exterior was, the simpler the interior. The suspension of the vehicle had to be sufficient. The stretcher resembled a campingbed that was supported by small springs. The right side of the ambulance had been fitted with venetian blinds.

26 Peugeot D3A (1957) built by Visser Bros., Leeuwarden, presumably for Verenigd Ziekenvervoer (VZA) in Amsterdam. The Visser brothers did not like the grille of this Peugeot and made more styled grilles for the succeeding Peugeot D4. The estimation of the labour costs of the grille had not been correct and therefore these ambulances had to be sold with a loss, for instance to VZA.

27 Studebaker (1957) of the Municipal Health Service (GGD) of Utrecht, built by Geesink, Weesp. These Studebakers had been very popular at many municipal services. Not only the Visser Bros. and Vermeulen built these ambulances, but apparently also Geesink. Geesink gained more reputation as bodyworker of garbage trucks.

28 Austin A152 (1958) of the Verenigd Ziekenvervoer Amsterdam (VZA), fitted up as ambulance by Visser Bros., Leeuwarden. Beside the much used Ford FK 1000s and the Commer 2500s, these courageous Austins were also used by some ambulance services.

29 Buick 700 Limited (1958) of the private Ziekenvervoer R. van de Beld & Zoon, Heerde. It was not until 1965 that Visser Bros. transformed this vehicle into an ambulance. The family-owned company Van de Beld originally used this vehicle as a wedding car, which was not uncommon for ambulances. Only three Buicks of this type have been sold in the Netherlands. Two of those were later fitted up as ambulance by Visser Bros.. One of these two burnt out completely in Leeuwarden. In 1958 the first blue flashing lights made their appearance on the ambulances.

30 Studebaker (1958) of the Municipal Health Service (GGD) of Nijmegen, built by Vermeulen, Haarlem. On this photograph the ambulance sniffs first the seawind in the dunes near Haarlem before it went to Nijmegen. This ambulance still has the transparent Red Cross beneath the blue flashing light. In many of the cases it is a red cross on a white surface. The use of that cross was not always legal as the red cross is only reserved for the Red Cross and thus only apt for ambulances of that same Red Cross. A white cross on a blue surface was not an uncommon view at that time. This cross originated in the association Het Witte Kruis (the White Cross) in North-Holland. This home-care association merged in the Groene Kruis (Green Cross). Until 1938 the White Cross had a committee, existing of doctors, who examined ambulances. If the ambulances passed the test they were allowed to transport members of the White Cross. The relation between the private ambulance service Ziekendienst Het Witte Kruis in The Hague and the association is not clear.

31   Cadillac series 60/62 (1959) of the Municipal Health Service (GGD) of Heerlen, presumably built by Akkermans, Oud Gastel. Next to the Cadillac stands a 1964 International Travelall, built by Visser Bros., Leeuwarden. The ambulance service of Heerlen had always been far ahead of its time. In the years when Jan ten Have was wardmaster and Balvert superintendent, nothing was a matter of course. Everything was researched seriously and creatively. New ideas that had been created were tried out. This picture shows the extraordinary American flashlights of aeroplanes. Even more particular is the rare extended roof of the American International. In Heerlen the first Cardulances were tried out in 1971 and 1972 (in association with Utrecht and Nijmegen). Officially the experiment failed, but as a result all the ambulances were fitted with coronary care units. At that time it was already possible to transmit wireless cardiac films to the cardiologist in the De Wever Hospital.

32   Volkswagen Minivan-ambulance (1961). Pon's Automobielhandel, Amersfoort imported these ambulances directly from West-Germany. These ambulances were solid, but in fact not suited as an ambulance. Yet, many were in Dutch service, as well for civilian purposes as for the Ministry of Defence, by the Royal Airforce with an extended roof. Many Volkswagens lost the side footboards because the step had not been pushed back and therefore was driven off.

33 Mercedes-Benz 300D (1961) of the private ambulance service Verenigd Ziekenvervoer Amsterdam (VZA), built by Visser Bros., Leeuwarden. This car was unique of its kind. Within VZA it was called 'The Golden Carriage' and was treated in such way. Only one driver was allowed to drive this ambulance. This vehicle was not used for every day work and was covered with a plastic dustsheet when not used.

34 Chevrolet Brookwood (1963) of the private firm Meeuwenoord & Zoon, Noordwijkerhout and built by Huiskamp, Winterswijk. This photo has been taken in 1973 in one of the crossroads of the Rijnsburgerweg near the former main entrance of the University Hospital of Leiden (AZL). It is not known whether or not this ambulance had been used by another ambulance service before entering service at Meeuwenoord. After ten years this vehicle still looked as good as new.

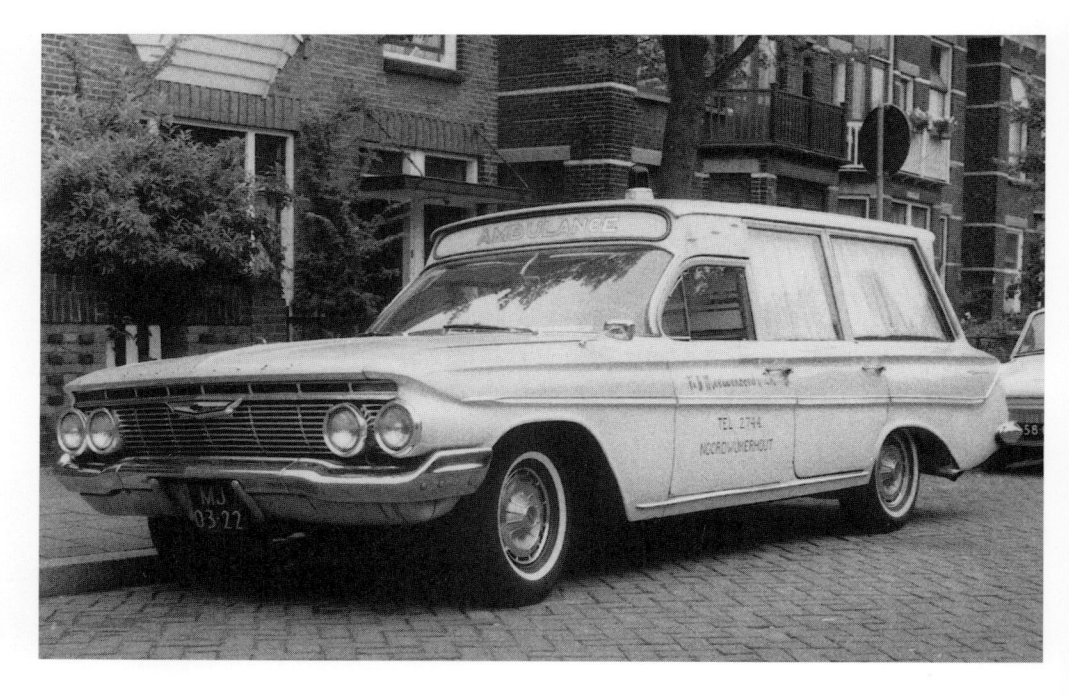

35 Chevrolet Apache 10 (approx. 1963) of the Municipal Health Service (GGD) of Haarlem, built by Vermeulen, Haarlem. Chevrolets of this type were in service by mainly municipal services for about ten years. Not only Vermeulen built ambulances on these Chevrolet chassis. Also Visser Bros. from Leeuwarden and Versteegen from 's-Hertogenbosch built on these chassis.

36 Opel-Kapitän (1963) of the St. Joseph Hospital, Heerlen, built by Miesen, Bonn (then: West-Germany) and delivered by NEDAM, Roermond. The importer's leaflet mentions that this 'technical miracle' (1962), 'the Opel-Kapitän ambulance car, of which already thousands are driving in- and outside Europe, is by far the most suitable vehicle to transport the seriously ill and injured'. A little bit further down you can read that 'the inner-room of the Kapitän offers place for two patients, it can be called spacious. So spacious, that even medical attention can be given during the journey to the hospital.' NEDAM recommended themselves as 'the only firm in the Netherlands with the disposal of an exhibition ambulancecar'. Miesen on the other hand obliged NEDAM always to have one car in stock. Within three years an odd 30 were sold.

37 Top: Ford Transit FK 1000 (1963) of the First Aid Service (Eerste Hulpdienst, EHD) of the Municipal Health Service (GGD) of Leiden, fitted up as ambulance by Visser Bros., Leeuwarden. This picture has been taken on the grounds of the fire station at Langebrug. Attendants had to participate in the 24-hours shift of the firemen as the drivers of the ambulances were firemen. They were used to drive large fire engines, which did not have power-assisted steering. The attendants were not always in for a ride with some of the firemen who had been posted on this FK 1000. Sometimes the attendants would sit in the back as the front of the vehicle is extremely close to their legs.

Under: Ford Transit FK 1000 (1963). Unbelievable but true: two patients could lie on a stretcher in these little Fords. Hardly any monitoring or treatment could take place. The only place to sit in such a case was in the chair next to the patients. A similar Ford was later sold to the first aid association in Hazerswoude-Rijndijk. There the Ford has served as an ambulance for another many years.

38 Commer (1964-1972) of the Municipal Health Service (GGD) of The Hague, delivered by car company Ten Hoeve, The Hague. After the Cadillac-ambulance era it was a huge step back to these Commers. They resemble the Ford FK 1000s and the Austin A152. In the leaflet stood 'These ambulances provide the comfort of a passenger car because of the independent front suspension system, the spacious cabin fitted for three persons, the easy to operate gear-lever, and a comprehensive set of instruments.' Nevertheless, many dangerous situations were created with these ambulances. They were not very stable. Even an fatal accident had occurred. At several ambulance services they were in great favour, probably because of the price tag. The private ambulance service VZA, Amsterdam also used these Commers.

39 Ford Country Sedan (1964) of the private firm Th. van der Laan, Nieuwkoop, built by Visser Bros., Leeuwarden. Van der Laan bought this ambulance from the Municipal Health Service (GGD) of Rotterdam. This municipal service always ordered their ambulances at Akkermans in Oud Gastel. The lower side was dark grey, the top white. This picture has been taken in 1972 at the corner of the Rijnsburgerweg and the Wassenaarseweg in Leiden. In the background stands building number 5 (outpatient department for internal diseases) of the University Hospital of Leiden (AZL).

40 Left: Ford Customline Sedan (1964) of the Municipal Health Service (GGD) of Breda, built by Akkermans, Oud Gastel. These cars were real battle ships as the bottom was red and the top was white. The angular shape was one of the exterior characteristics of Akkermans. The interior was very sober. They were only suited for the transportation of the sick, despite the spacious cabin.

Right: Ford Custom Sedan (1964). Special of this ambulance was that the entrance on the right of the vehicle could be moved upwards so that patients in a chair could be lifted in.

41 Left: Opel Blitz (1964) of the Municipal Health Service (GGD) of Delft, built by Miesen from Bonn and delivered by NEDAM, Roermond. The original intention for this ambulance was to assist at large accidents on national highway 13 (The Hague-Delft-Rotterdam). At that time it was a dual carriageway, separated by a hedge in the middle and no safety central reserve.

Right: Opel Blitz (1964). This Opel is referred to as the vehicle in case of big accidents and had room for four stretchers in a permanent formation. It still was no care unit, but more a 'pick up and go' ambulance.

42 Cadillac (1965) of the Municipal Health Service (GGD) of The Hague, built by Visser Bros., Leeuwarden. For the larger communities in the Netherlands it were wealthy years, in which they could afford to purchase the expensive Cadillacs. The Cadillacs provided a magnificent place to work in. The suspensions were superb. Up to three stretchers could be placed in the cabin. The ambulances at the service of The Hague were recognisable by the letters on the vehicles. This particular Cadillac had the letter Q and had the nickname 'the Golden Carriage' (de 'Gouden Koets', called for the carriage of Queen Beatrix at the opening of the governmental year). This Cadillac was the longest running Cadillac as an ambulance until it was also replaced by a Mercedes-Benz.

43   Chevrolet Bel Air (1965) of the private ambulance service De Jong Bros. from Leiden, built by Visser Bros., Leeuwarden. One son of each of the founders, who were brothers, took over this renowned company, which was situated on 164, Hogewoerd. Two other sons took over the private ambulance service Ziekendienst Het Witte Kruis, The Hague. The company in Leiden also performed weddings and funerals. For these purposes they used original wedding and funeral horse-drawn carriages. All the ambulances were cleaned on Saturday. On that day also the stables were mucked out. The proof of this fact can be seen behind the ambulance on the picture.

44 Mercedes-Benz 190 (1965) of the private ambulance service Eigenbrood from Lisse, built by Binz, Lorch (West Germany). This type with extended roof was called Europ 1100. Binz had already built a large number of ambulances on chassis of Mercedes-Benz passenger cars for as well the national as the international market. This type was the first that had been imported in the Netherlands. Next to the imported Mercedes-Benz ambulances from Binz and Miesen, also the Visser Bros. from Leeuwarden designed and built ambulances on Mercedes-Benz chassis. The most striking feature is the Maltese Cross in the roof transparant, which raises the suspicion that this ambulance had been imported from West Germany. The registration number dates from 1968. The Maltese Cross also indicates a first aid service. This picture had been taken on the grounds of the University Hospital of Leiden (AZL) in 1973. At that time this ambulance was already eight years of age, which was not a surprise.

45 Volkswagen Minivan (1966) of the Juliana Hospital, Apeldoorn and imported by the Volkswagen-importer Pon, Amersfoort. This ambulance also has the Maltese Cross in the roof transparent. This Volkswagen dates from the time that the two side doors on the right were opened to both sides. Later it would become only a sliding door on the right side. These ambulances were not very stabile on the road. In Delft a similar kind of ambulance had been launched when a student, driving a Harley-Davidson motor-cycle, did not gave way to the ambulance and drove onto its rear tire.

46   Left: Mercedes-Benz 230 Lang (1966) of the Municipal Health Service (GGD) of The Hague, built by Visser Bros., Leeuwarden. This type was the foundation of the preference for Mercedes-Benz ambulances by many ambulance services. Every male nurse had his own first aid kit, with bandages and first aid material and two small bottles. One bottle contained ammonia, and the other brandy. The ammonia was used for patients who did not really fell faint; they could not offer resistance to the ammonia. The brandy was used to recuperate the patient with a more serious faint. Some substitute male nurses from the University Hospital in Leiden had to fill the last bottle more often then their colleagues. Maybe it was because they were students!

Right: Mercedes-Benz 230 Lang (1966). Three patients could be put in on three stretchers. The equipment was still very concise. The lock of the stretchers was not always as safe as it should have been. One incident took place when driving to the Municipal Hospital, located at Zuidwal, in the famous Boekhorststraat in The Hague. The backdoor sprang open and the stretcher with patient was launched in the middle of the street!

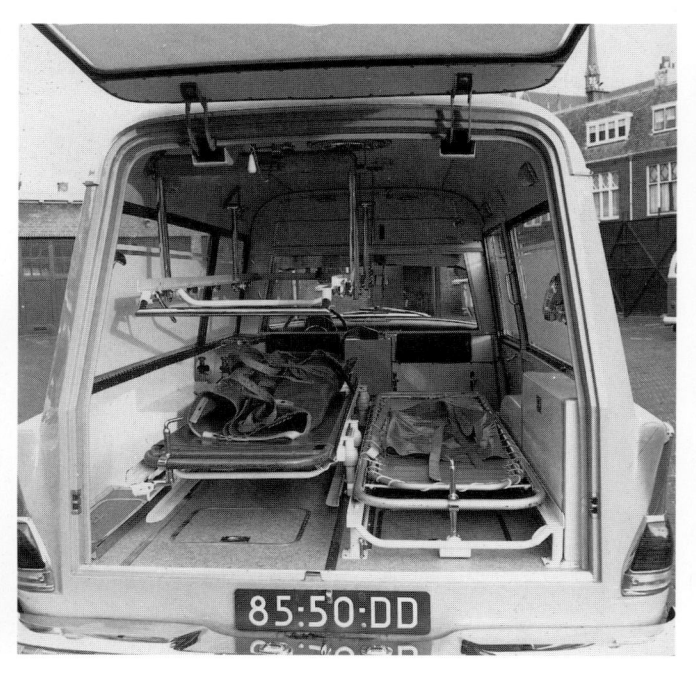

47   Cadillac (1966) of the private ambulance service R.v.d.Beld & Zoon, Heerde. This ambulance was fitted up by Visser Bros., Leeuwarden. Van de Beld originally used this car for wedding purposes during the first two years before it was transformed into an ambulance. Ambulances with an extended Cadillac chassis served for many years at the larger community services. This ambulance on the other hand is built on a short, thus normal chassis. Probably this Cadillac was the only one of its kind in the Netherlands.

48 Mercedes-Benz 230 Binz Europ 1100 Lang (1967) of the Municipal Health Service (GGD) of Heerlen, built by Binz, Lorch (West Germany). A solid ambulance with two unusual American, blue flashing lights from an aeroplane, which gave a bulk of light. The immense ambulance transparent had been put on by the GGD themselves. The spare wheel of this type of ambulance had been placed in a compartment above the right rear tyre.

49   Opel Kapitän (1967) imported from West Germany by NEDAM, Roermond. Especially many ambulance services and hospitals, which provided the transportation of the seriously ill and injured, used these Opels frequently. The search light on the left, above the windscreen was a standard feature on ambulances.

50 Chevrolet C/10 (1967-1968) of the private ambulance service Grave from Grave, built by Versteegen, 's-Hertogenbosch. Ambulances built on Chevrolet chassis were very popular in those years. Versteegen did not only build large backdoors in the ambulances, but also built similar windows in the sides and in the backdoors. The spare tyre had been placed in a compartment behind the front door on the left. When you look through the window of the ambulance you can see the oxygen apparatus and the respirator balloon. These two features were the first signs of the introduction of immediate emergency treatment in the transportation of the seriously ill and injured.

51 Peugeot J7 (1967-1968) of the First Aid Service (EHD) of the Municipal Health Service (GGD) of Leiden, fitted up as ambulance by Visser Bros., Leeuwarden. This picture has been taken in the courtyard of the fire department at Langebrug. These Peugeots replaced the Ford FK1000, which had been used for many years. In these years the EHD only provided the emergency transportation in the region of Leiden. The EHD wanted to have an ambulance in which real emergency treatment would be possible. By replacing the old Fords, the EHD anticipated to the Ambulance service act (Wet Ambulancevervoer). The spacious workrooms in the Peugeots realised the emergency treatment. The Peugeots also gained some popularity in the Netherlands after their success in France as an ambulance for emergency transportation and emergency treatment.

It was possible to bring the stretcher up to a height so that on the left, on the right, and at the head-end of a patient treatment could be given. The second stretcher had been placed in the ambulance as required by law, though its function was merely as an emergency stretcher. The chest of drawers on the left side contained a wooden Verhees scoop-stretcher that was replaced by an aluminium one made by Ferno-Washington. Besides the blue flashing lights, these ambulances were also fitted with and an extensible yellow hazard flasher on the back.

52 Chevrolet Biscayn (1968) of the private ambulance service Ziekendienst Het Witte Kruis B.V., The Hague. This picture has been taken in front of the outpatient department of the municipal Leyenburg Hospital. Het Witte Kruis did not take part at that time in the transportation of the seriously ill and injured or the emergency transportation. The company fitted up American estate cars for the transportation of the ill themselves. These cars came from the RIVA (General Motors dealer) in The Hague. These ambulances were more or less the precursors of the auxiliary ambulances or recumbent taxis with which the company had been allowed officially to experiment with some twenty-five years later.

53 Left: Chevrolet Malibu (1968) of car company Felman from Hellevoetsluis and built by Akkermans, Oud Gastel. The original car had only been provided with an extended roof. Even the lid of the boot with the revolving window had not been adjusted. The patient could assume a sitting posture, but when the patient was shoved in or out he had to draw in his head (right picture).

54 Mercedes-Benz 230-lang (1968) of the private ambulance service De Jong Bros., from Leiden, built by Visser Bros., Leeuwarden. At the beginning the Chevrolet ambulance was in great demand during the two generations of De Jong as well in Leiden as in The Hague. These two companies on the other hand became loyal Mercedes-Benz users. In 1966 Dijkstra, surgeon in Woerden, made clear that a patient had to be transported between the axles of the wheels. The brochures of the extended versions of this Mercedes-Benz and of the Opel-Kapitän made reference to this aspect. By hook or by crook a patient could be transported between the axles (the stretcher had to be shoved in all the way up to the partition-wall). This was not possible of course.

55 Plymouth Suburban (1968) of the private ambulance service A. Niemansverdriet from Spijkenisse, built by Akkermans, Oud Gastel. This Plymouth was a unique vehicle with a distinguishing exterior.

56 Mercedes-Benz 408 (1968-1969) of the Central Board of the Dutch Red Cross (NRK), fitted up as ambulance by Visser Bros., Leeuwarden. The NRK had more Mercedes ambulances like this one. These were brought into action as motorway ambulances, a special project of the NRK, started in 1966 by demand of the General Traffic Service (Algemene Verkeersdienst). Not only these Mercedes were bought or donated to the NRK, but also additional training was given to many of the hundreds of Red Cross volunteers. A maximum of two patients on a stretcher could be transported in this ambulance. There were several ambulance-stands. The ambulance-stand on the picture is in the vicinity of the no more existing Shell-petrol station at Gouwebrug (Gouda), near national highway 12. These ambulances were only used during risk-bearing weekends and public holidays.

57 Chevrolet C/10 (1969) of the Municipal Health Service of Amsterdam and presumably built by Akkermans, Oud Gastel. The municipal service of Amsterdam used to have tens of similar ambulances as part of its fleet. These vehicles replaced the Citroën HY. These Citroëns had replaced the incomparable Cadillacs. This picture has been taken on the grounds of the University Hospital of Leiden (AZL) in front of the paediatric cardiology, opposite the class pavilion of internal medicine.

58 Mercedes-Benz 220D-lang (1969) of the private ambulance service De Jong Bros. from Leiden, built by Visser Bros., Leeuwarden. The colour of the ambulance was mainly grey and had a green band, what was a characteristic feature for the company. In the roof were the so-called mountain windows (bergramen). These windows especially occurred in coaches so that the passengers were able to look through the window and enjoy the mountainous scenery. The windows in the ambulances had the same purpose. These ambulances were especially used to transport patients who had been hospitalised in the University Hospital of Leiden (AZL) and to make outings with chronically sick patients as if it was a coach.

59   Fiat 238 (1970) of the rural hospital Prinses Beatrix from Gorinchem and fitted up as ambulance by Versteegen, 's-Hertogenbosch. The manufacturer of these Fiats could supply the exterior with an extended roof. The roof was made out of synthetic material, just like the roofs of the Mercedes and Chevrolet with a Dutch exterior. The reason for the synthetic material was to diminish the weight of the ambulances. The exterior suggests a spacious cabin, the truth is on the other hand that the length of the interior was hardly long enough for a Dutchman, who was lying on a stretcher. There were no cheaper ambulances that could be delivered in those days.

60   Fiat 238 (1970) of the Dutch Red Cross and fitted up as ambulance by Akkermans, Oud Gastel. The Red Cross had a large number of these so-called PAM-ambulances at their disposal. PAM is the abbreviation of Personnel-Ambulance-Materials. These ambulances were multifunctional, intended for the transportation of personnel and equipment to the place of the calamity, and of course for the transportation of the victims to the hospital.

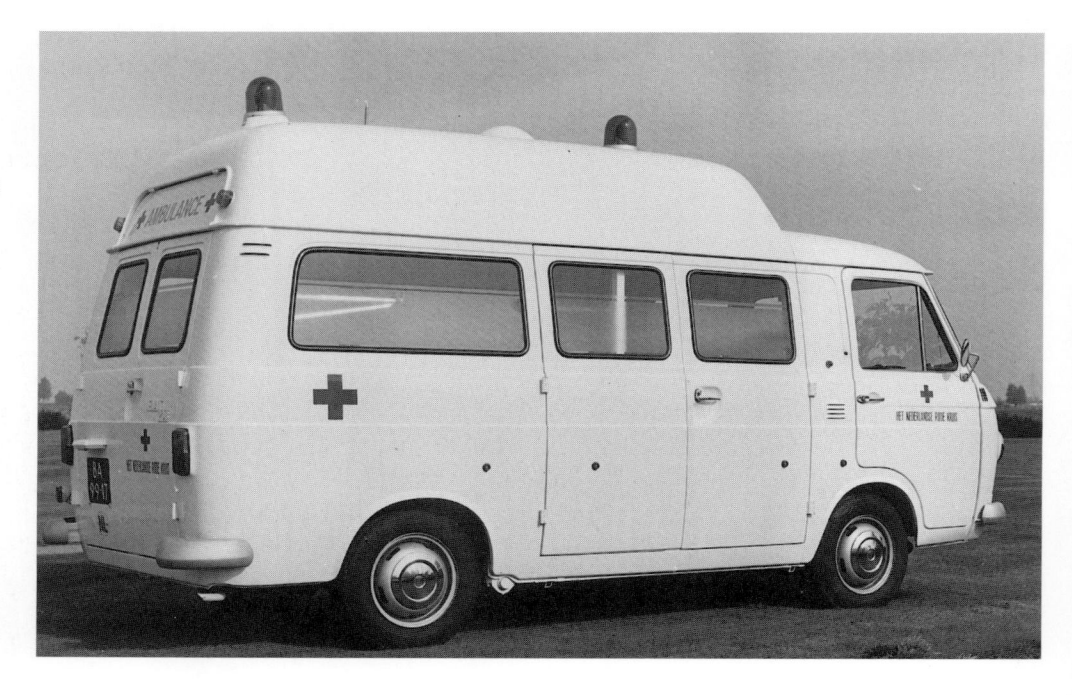

61   Mercedes-Benz 230-lang (1970) of the Municipal Health Service (GGD) of Rotterdam, built by Akkermans, Oud Gastel. The way Akkermans built the ambulances for their regular clients remained the same over the years; a high roof, angular, and firm. Even the front doors were fitted with a small window on the upper side.

62 Mercedes-Benz Europ 1200L (1970) of the Municipal Health Service of Velsen, built by Binz, Lorch (former West Germany). The indication '1200' meant that the ambulance had an extra extended roof. De 'L' indicated a 25,6 inch extended chassis. The German brochures referred to the cabin as 'der Krankenraum wird zum Behandlungsraum' (the patient-accommodation becomes a therapy theatre). Binz reacted to the only just rediscovered mouth-to-mouth resuscitation and the external cardiac massage by remarking that 'Alles in allem ein Fahrzeug ist, das auch aktuellen medizinischen Forderungen zur Sofortbehandlung Unfallverletzter gerecht wird' (Altogether an ambulance that offers new possibilities for emergency therapy to accident casualties). The service from Velsen had two more accessories installed, namely a special grill on the right of the large grille for the sound of the horn, and a loudhailer on the left of the grille so that fellow road-users could be called upon to clear the way if necessary. The ambulances of the GGD were operated by the private ambulance service Kloosterhuis from IJmuiden.

63 Top: Mercedes-Benz 408 (1970) of the private ambulance service L. Hoek from Woerden, fitted up as ambulance by Visser Bros., Leeuwarden. It probably was no coincidence that such an ambulance was in service in Woerden. In 1962 a serious train-crash occurred in Harmelen, near Woerden. The problem that arose was there was no Dutch standard concerning the stretchers. The result was that patients could not be transported to hospitals on a stretcher which belonged to an another ambulance. This accident was the reason that the surgeon from Woerden, Dijkstra, in the 1960s indulged himself in the demands that had to be set for the transportation of the victims of serious accidents.

Bottom: Mercedes-Benz 408 (1970). One of the demands that the surgeon Dijkstra set was more workspace in the patientaccommodation. A second demand was that the patient could be given treatment from any side. Not all the demands were realised though. It remained a question of doing something for the patient before and during the transportation to the hospital, or pick-up and go. The three stretchers in the picture indicate the last strategy.

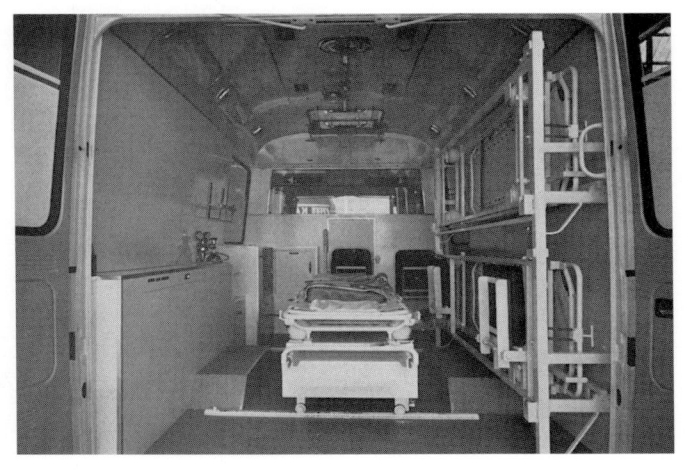

64 Opel-Kapitän Bonna (1971) of the Municipal Health Service (GGD) of Ede, built by Miesen, Bonn (former West Germany) and delivered by NEDAM, Roermond. The stretcher in this extended Opel could be shoved in up to the partition wall, so the patient would lie between the axles of the wheels. More important is that the patient could be put in the Trendelenburg-posture (head lower compared to the pelvis, like advised to do so for patients with serious haemorrhage). The boot existed of two parts; the lower lid had been fitted with a revolving tray (the stretcher would be pulled out and turned on the lower lid, so that the two bearers strain themselves in lifting the stretcher. The GGD of Ede was one of the first ambulance services of those years in the Netherlands to drive around in dark yellow ambulances out of safety measures.

65 Bedford CF (1975) of the Municipal Health Service of 's-Hertogenbosch, of which the outside is build by Binz, Lorch (former West-Germany) and the inside by the public works department from 's-Hertogenbosch. In Great Britain this van was a very popular chassis and body to built ambulances on. Presumably this is the only specimen on Dutch soil. Because of the extended frame the side door, which gave access to the cabin, opened all the way to the back-axle. The step could not be pushed in so it frequently occurred that the step was driven off.

66 Chevrolet Nomad (1971) of the private ambulance service Ziekendienst Het Witte Kruis, The Hague, photographed in front of the outpatient department of the municipal Leyenburg Hospital. Het Witte Kruis themselves fitted up this ambulance and was only used for the transportation of the non-emergency cases. It was the permanent vehicle driven by the current managing director, Ruud de Jong. The oxygen flask lay behind the driver's seat and was given through the partition wall if necessary.

67 Chevrolet C/10 (1972) of the Municipal Health Service of Haarlem, built by Visser Bros., Leeuwarden. Remarkable is the extremely extended door in the front (with extended window). This ambulance has been photographed at the back of the outpatient department building of internal diseases of the University Hospital (AZL) at the Wassenaarseweg in Leiden.

68 SA or SAM (1972). No-one has yet dared to put this Emergency Ambulance (SpoedAMbulance) on the road, designed by Jansdaal. Without any restrictions Jansdaal designed this ambulance as his graduate assignment at the University (former Technische Hogeschool) of Delft. His opinion was that up to that moment no ambulance met some basic requirements concerning the treatment and the transportation of the seriously ill and injured. In addition he could meet the requirements which had been formulated by the surgeon from Woerden, Dijkstra. The most important demands were that the patient had to lie as close to the ground as possible and between the axles of the vehicle, with his head pointing to the backdoor. The attendant had to be able to access the patient from any side of the cabin. But most important was the space near the head of the patient, as there the most life saving performances would be given. A full headroom had to be realised for the attendants (6ft3inch instead of the demanded 5ft3inches by the Ambulance Service Act). With this graduate assignment Jansdaal obtained his engineering title in technical and industrial design with first class honours (cum laude).

69 Chevrolet Impala (1973) of the private ambulance service Ziekendienst Het Witte Kruis from The Hague, built by Akkermans, Oud Gastel. With this Impala Het Witte Kruis was bound to become a well-noted ambulance service, here photographed in front of the First Aid entrance of the Red Cross Hospital in The Hague located at Sportlaan. But for an equal place within the ambulance and emergency services in The Hague years of struggle with the Municipal Health Service (GGD) were to be followed. The repatriation transport of the Tourist Organisation ANWB (the Dutch AA) provided an important impulse for the professionalization and the aim of a worthy place for this ambulance service in the region of The Hague.

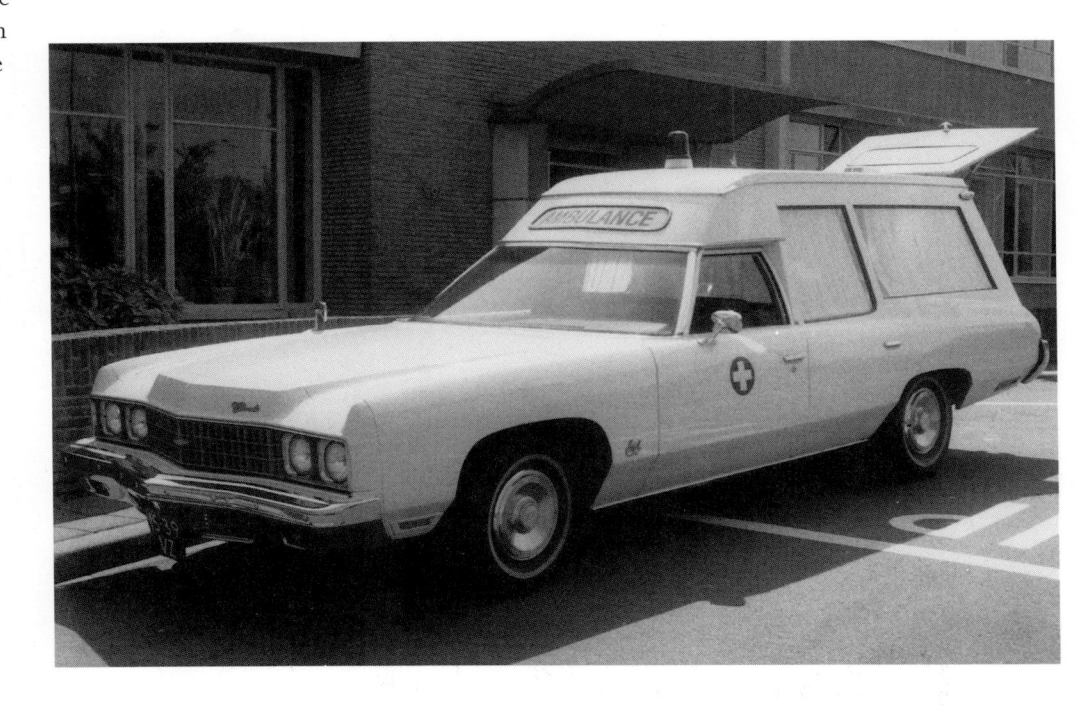

70 Dodge B200 Tradesman (1974). After the Ambulance Service Act came into force (1971) and more attention was given to the real needs of the patient who needed immediate emergency attention, ambulances were presented on the chassis of a Dodge B200 Tradesman by car company Ten Hoeve from The Hague. Ten Hoeve was also the company who introduced the Commers in 1960. As well Akkermans from Oud Gastel (top) as Visser Bros. (bottom) from Leeuwarden were invited to fit up these ambulances. Akkermans did so by extending the roof across the whole length of the vehicle, and by putting in a split door, which opened to the side in order to access the cabin. Visser Bros. placed an extended roof made of synthetics with on the front a kind of terrace for the flashing lights. In order to access the cabin they placed a sliding door. The Municipal Health Service (GGD) of Amsterdam saw the Dodge as the replacement of their fleet of Chevrolets C/10, but the newcomer became a no-go when up to three times in action the Dodge fell on its side.

71 Hanomag F20 (1974) of the Municipal Health Service (GGD) of Rotterdam, built by Akkermans, Oud Gastel. The headline of the article in the newspaper NRC (Nieuwe Rotterdamse Courant of January 24th, 1974) said that the GGD tried out their new ambulances. The GGD of Rotterdam had been allowed to put three of these new ambulances to the test. They were built on the van-chassis and fitted with the so-called 'vleugelbrancard' (anti-vibration and anti-shock floating stretcher). Ambulance services wanted a more spacious cabin, which could only be built on these chassis. The suspension of these vans on the other hand was stiff and not really suited to transport patients. At the University (former Technische Hogeschool) of Delft the 'vleugelbrancard' had been developed. This stretcher was able to absorb the shocks and vibrations. These stretchers therefore eliminated most of the problems for using a delivery van frame. The 'vleugelbrancard' had been designed by LauraViCo Vibration Control, a company that emerged from the former coal mine Laura and with the exploitation of new activities tried to limit the loss of jobs in the mining sector in Limburg.

72 Volkswagen Minivan (1974). Presented by Volkswagen at the car show Personenwagen-RAI of 1974 in Amsterdam, built by Visser Bros., Leeuwarden. An old design resulted in a new ambulance. A synthetic, extended roof had to increase the working-space. The sliding door made access to the patientaccommodation easier. The patient not only still lay above the rear axle, but also above the engine, which is a normal feature of Volkswagen. Neither the previous ambulance, the Hanomag, nor this Volkswagen would survive in the Netherlands as an ambulance.

73 Opel Admiral Bonna (1974), built by Miesen, Bonn (former West Germany) and delivered by NEDAM, Roermond. On the picture this ambulance stands at the car show Personenauto-RAI of 1974 in Amsterdam. Even though a few had been sold to some services, it never had the success of its rival from Mercedes-Benz. Opel tried it again a few years later with a Senator (1980s) and after that an Omega (1990s).

74 Peugeot J7 (1974) of the local council Hazerswoude, built by Visser Bros., Leeuwarden. This ambulance was run by the first aid association of the village and was the replacement of a second-hand Ford FK1000 of the first aid association The Orange Cross (Het Oranje Kruis) of the Municipal Health Service of Leiden. It was one of the first real emergency ambulances in the Netherlands. The first aid association and the author of this book had been allowed to develop this ambulance. An extensive program had been developed with detailed designs of the interior as well as for the exterior. The experiences of the Peugeots from Leiden were thankfully used. Even a few ideas of the Spoedambulance, which had been designed at the University (former Technische Hogeschool) of Delft, were used. This ambulance had been fitted with the floating Laura-stretcher and a revised dentist's chair so that people who fell faint could be placed in a recumbent position as soon as possible. In later years blue flashers would be placed in the front. This ambulance was presented by Visser Bros. in 1974 at the car show Personenauto-RAI in Amsterdam. The ambulance was presented again at the first resuscitation day of the VVAA (Dutch Association of Physicians) in the Jaarbeursgebouw, Utrecht. The wife of the author made the first unofficial ride in this ambulance, when she had to give birth to their first daughter.

75 Mercedes-Benz 230 Binz Europ Lang (1975). Up to this day in the Netherlands the extended Mercedes-Benz passenger car chassis and the Chevrolet-van chassis are the most popular frames to built ambulances on. Every coachbuilder, as well inside as outside the Netherlands, has successfully realised its own design.

With thanks to Mrs.E.Waldeck-Koster, Ir.J.P.Koster, Henri Geurts, Thijs Gras, Jan ten Have, Ruud de Jong, and Bert Visser.

Much attention has been given to the accuracy of the information. Nevertheless, mistakes or inaccuracies can occur in the texts. Corrections or additions are extremely appreciated and can be sent to K.J.J.Waldeck, M.D., Ph.D., 'de ijsvogel', Commissieweg 1, 7957 NC de Wijk, the Netherlands. Telephone ++.31 522.443.019, fax ++ 31 522.443.064, e-mail: kwaldeck@worldonline.nl

Used sources:

- Gras, Th. 'Met zorg op weg' Verenigd Ziekenvervoer Amsterdam 1957-1997. Een eeuw particulierziekenvervoer in Amsterdam. Amsterdam: HHS Uitgeverij, 1997.
- Hoving, P.G. 'Nederlandse ambulances 1908-1970' De Ambulance Nijmegen, 1982.
- Hoving, P.G. '175 Jaar Carrosserie Akkermans' De Ambulance 7(1986):4 (21-25).
- Keuzenkamp, J.H. 'Gemeentewapens in Nederland 1914-1989 naar het officiële register van de Hoge Raad van Adel' The Hague: VNG-Uitgeverij, 1989.
- Smit, W.B. 'Organisatie en werkwijze van de ongevallendienst te Haarlem' Het Reddingwezen 40(1951):2 (35-36).
- Vanderveen, Bart H. 'American Cars of the 1940s' London: Frederick Warne & Co Ltd (Olyslager Auto Library), 1972.
- Vanderveen, Bart H. 'American Cars of the 1950s' London: Frederick Warne & Co Ltd (Olyslager Auto Library), 1973.
- Vanderveen, Bart H. 'British Cars of the Early Fifties 1950-1954' London: Frederick Warne & Co Ltd (Olyslager Auto Library), 1975.
- Zuijlen, Rob van. 'Van handkar naar ambulance' Purmerend, 1996.

- Several brochures and product information leaflets from bodywork companies and importers.
- Several volumes of the magazine 'De Ambulance' (1979-1993), and 'Nederlands Tijdschrift voor Spoedeisende medische hulpverlening en Rampengeneeskunde' (Dutch Journal of Emergency Services and Disaster Medicine) (1994+).
- Several memorial volumes of hospitals.

The National First Aid and Ambulance museum

The National First Aid and Ambulance Museum is located at De Papierbaan in Winschoten in one of the buildings of car company De Grooth (importer of American (Superior) and German (Binz) ambulances and provides the (private) ambulance service for the region of Winschoten in addition. The collection is part of a foundation (1988), but is the continuation of the private collection of C.A.E.Volckmann, M.D., who used to work for the Municipal Health Service of Groningen. The objective of the museum is to preserve and to document, and to make access to the material easier for researchers and the public. You can look at the material that had been used and is still being used in the first aid and ambulance services. The museum has thirteen ambulances as part of their collection of which regrettable a few are still in a deposit. You can visit the museum daily by request from 9am till 4pm. (telephone ++31.597.422.000.

The Dutch Ambulance Archives

Four people founded this archives in the spring of 1998. The objective of this initiative is to make an inventory of everything that has a recollection with ambulances and the care for them, and if necessary to preserve the material. Thereby the initiators want to stimulate the historical research or do the research themselves. At present a probe has been set up into all ambulance services, which have existed in the Netherlands (Thijs Gras) and all ambulances cars that have ever been on the road in Holland (Hans Waldeck). This initiative has sought affiliation with the stocktaking of medical-historical research in the Netherlands by the Huizinga Institute, Amsterdam, and the Rudolf-Agricola Institute.

Reactions and information are welcomed by the initiators:

Henri Geurts, P.O.Box 72, 5430 AB   Cuyk, the Netherlands.
Tel/Fax:++.31.485.321.708.
Thijs Gras, Van Walbeeckstraat 251, 1058 CG   Amsterdam, the Netherlands.
Tel.++.31.20.6166.347 - E-mail : t.gras@hetnet.nl
Piet Hoving, c/o HHS Uitgeverij, P.O.Box 150, 5360 AD   Grave, the Netherlands.
Hans Waldeck, 'de ijsvogel', Commissieweg 1, 7957 NC   de Wijk, the Netherlands. Tel.++.31.522.443.019, Fax.++.31.522.443.064.
E-mail : kwaldeck@worldonline.nl